STARTING OUT

I0803711

# baby BATS

MEG GREVE

CREATIVE EDUCATION • CREATIVE PAPERBACKS

# CONT

# ENTS

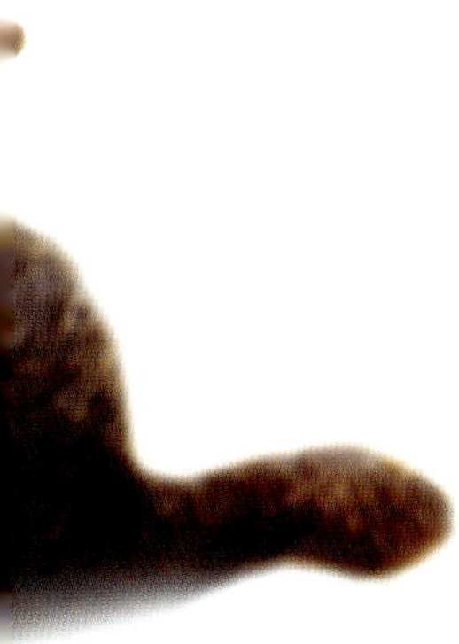

# I AM A PUP.

I am a baby bat.

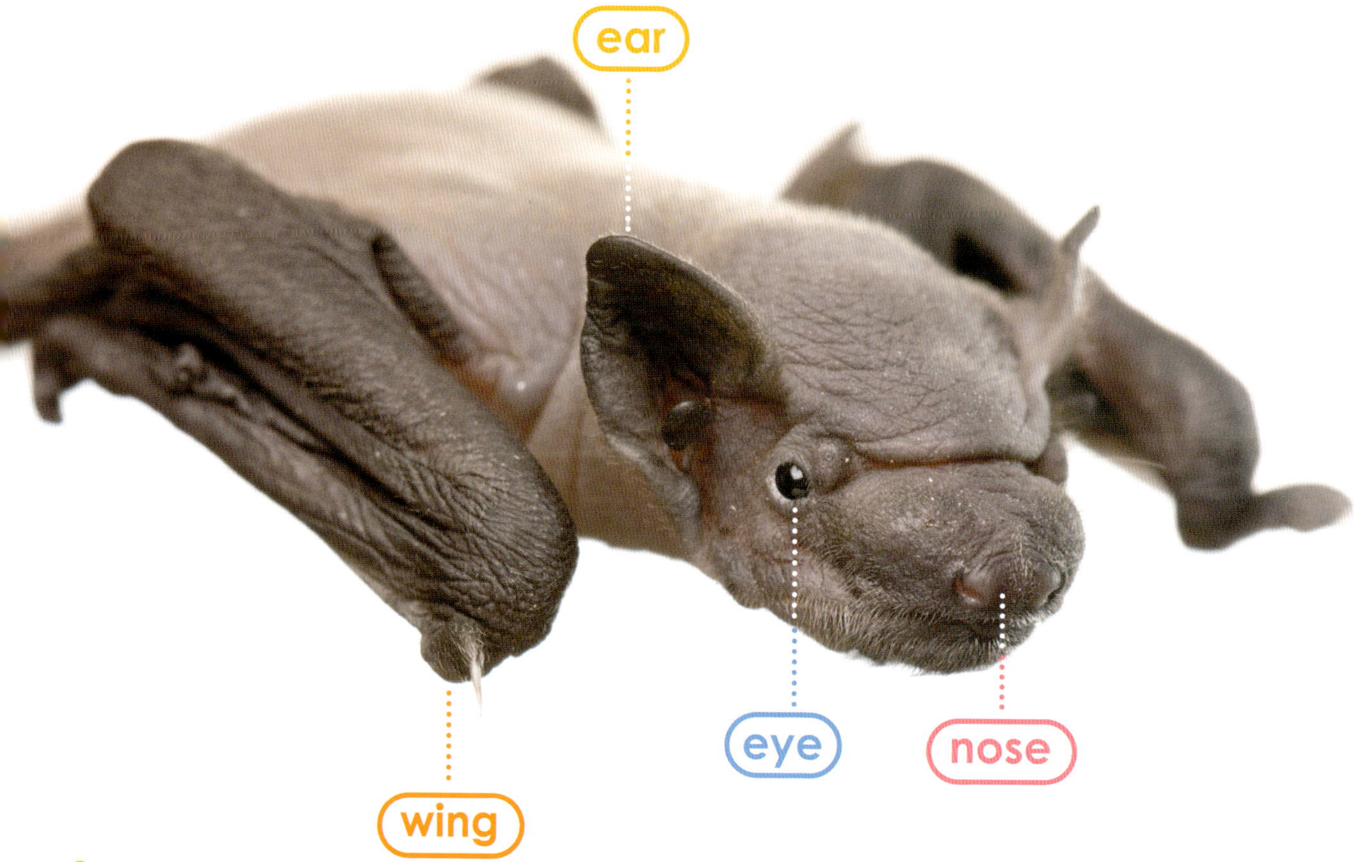

I was born
with no hair.
It will grow
later.

I use my claws to hold on to my mom.

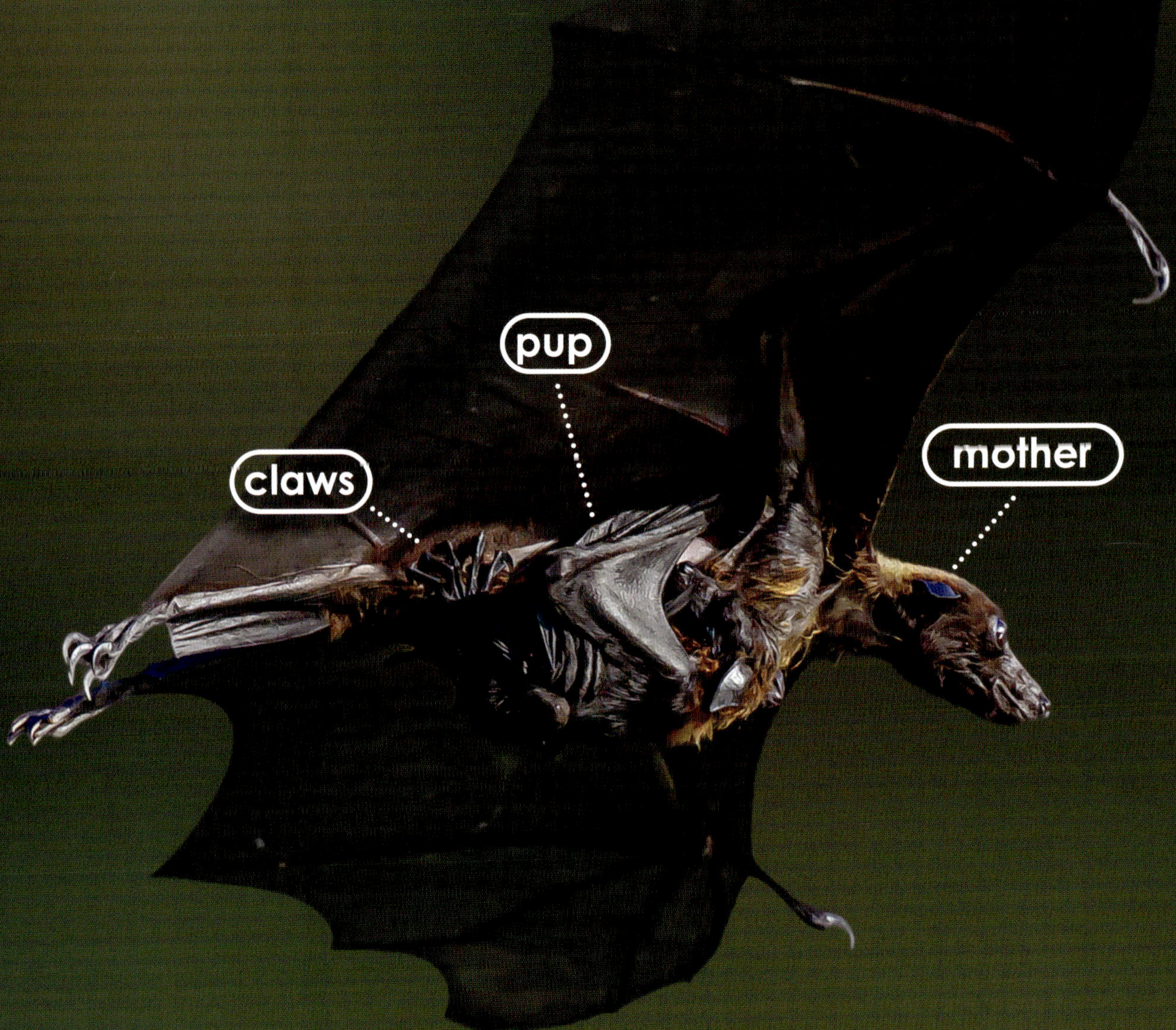

I drink my mom's milk.

**I live with many** bats.

We are a colony.

We sleep all day.

We hang upside-down in a cave.

I fly, but I am not a bird.

After six weeks, I can fly.
I can hunt for insects.

Now I can take care of myself!

# SPEAK AND LISTEN

# RRPP!

Can you speak like a pup?

Bat pups chirp and squeak.

Listen to these sounds:

https://www.youtube.com/watch?v=5ZhuDsdSjgs

Now it is your turn!

# PUP WORDS

**cave:** an opening under the ground

**claws:** sharp nails on a bat's feet

**colony:** a group of bats that live together

**insects:** bugs

# READING CORNER

Animal Reads. *All Things Bats for Kids*. Atlanta, Geor.: Admore Publishing, 2022.

Jenner, Caryn. *All about Bats*. New York: DK Children, 2017.

Pajaro, Rosie. *PBS Kids Bats!* Rolling Meadows, Ill.: Cottage Door Press, 2021.

# INDEX



**PUBLISHED BY CREATIVE EDUCATION AND CREATIVE PAPERBACKS**
P.O. Box 227, Mankato, Minnesota 56002
Creative Education and Creative Paperbacks are imprints of The Creative Company
www.thecreativecompany.us



**LIBRARY OF CONGRESS CATALOGING-IN-PUBLICATION DATA**
Names: Greve, Meg, author.
Title: Baby bats / Meg Greve.
Description: Mankato, Minnesota : Creative Education and Creative Paperbacks, [2025] | Series: Starting out | Includes bibliographical references and index. | Audience: Ages 4-7 | Audience: Grades K-1 | Summary: "Introduce beginning readers to the winged world of baby bats with this life science starter. Includes photos, a labeled animal diagram, "Make a Noise" section, glossary, and further resources"-- Provided by publisher.
Identifiers: LCCN 2024014694 (print) | LCCN 2024014695 (ebook) | ISBN 9798889892069 (library binding) | ISBN 9781682775714 (paperback) | ISBN 9798889893165 (ebook)
Subjects: LCSH: Bats--Juvenile literature. | Bats--Infancy--Juvenile literature.
Classification: LCC QL737.C5 G693 2025 (print) | LCC QL737.C5 (ebook) | DDC 599.413/92--dc23/eng/20240419
LC record available at https://lccn.loc.gov/2024014694
LC ebook record available at https://lccn.loc.gov/2024014695

**DESIGN AND PRODUCTION**
Design by Rhea Magaro
Production by Beeline Media and Design, Inc.
Art direction by Tom Morgan

**PHOTOGRAPHS** by Alamy Stock Photo/cbstockfoto, 7, Goss Images, 11, MYN / Paul van Hoof, 4; Dreamstime/Picstudio, 2, 12; Freep!k/ jl.lago, cover; Getty Images/peters99, 10; Shutterstock/ Bohbeh, 13, Goensaja, 5, I Wayan Sumatika, 9, Nico Faramaz, 14, Nitin Chandra, 6, pzAxe, 8

Printed in China